δ Delta

Single and Multiple-Digit Division

Tests

1-888-854-MATH (6284) - mathusee.com
sales@mathusee.com

Delta Tests: Single and Multiple-Digit Division

©2012 Math-U-See, Inc.
Published and distributed by Demme Learning

mathusee.com

1-888-854-6284 or +1 717-283-1448 | demmelearning.com
Lancaster, Pennsylvania USA

ISBN 978-1-60826-075-1
Revision Code 1118-E

Printed in the United States of America by Command Companies
4 5 6 7 8 9 10

For information regarding CPSIA on this printed material call: 1-888-854-6284
and provide reference #1118-03172021

Fill in the parentheses with the factors and write the product in the oval. Then write the problem under the rectangle.

1.

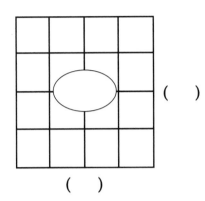

( )

( )

____ × ____ = ____

2.

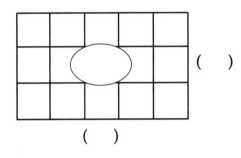

( )

( )

____ · ____ = ____

____ · ____ = ____

Solve for the unknown.

3. 3X = 27

4. 8Y = 64

5. 4Q = 20

6. 6B = 6

7. 5D = 45

8. 6F = 24

9. 10R = 100

10. 2H = 14

Find the area.

11.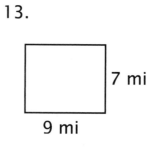

5 ft

5 ft

Area = _____

12.

10"

8"

Area = _____

13.

7 mi

9 mi

Area = _____

14. Dan earns $9 an hour. He earned $90 last week. How many hours did he work?

15. A room measures eight feet by nine feet. What is the area of the room?

Divide.

1. $8 \div 1 =$ _____

2. $18 \div 2 =$ _____

3. $6 \div 2 =$ _____

4. $4 \div 1 =$ _____

5. $10 \div 2 =$ _____

6. $8 \div 2 =$ _____

7. $\dfrac{2}{2} =$ _____

8. $\dfrac{7}{1} =$ _____

9. $\dfrac{4}{2} =$ _____

Solve for the **unknown**.

10. $5X = 15$

11. $5R = 25$

12. $2B = 20$

13. $10Y = 60$

Multiply.

14.   8
    × 5

15.   1 0
    × 7

16. 4 · 5 =_____

17. (10)(3) =_____

Find the area. Remember to label your answer correctly.

18.

8"

10"

A = _____

19. Madison divided 12 gumdrops evenly between herself and a friend. How many gumdrops did her friend receive?

20. A pie is cut into eight pieces. There are eight people around the dinner table. How many pieces of pie can each person have?

Divide.

1. $10\overline{)20}$

2. $10\overline{)80}$

3. $10\overline{)60}$

4. $10\overline{)90}$

5. $1\overline{)8}$

6. $2\overline{)14}$

7. $2\overline{)16}$

8. $1\overline{)5}$

9. $6 \div 2 = $ _____

10. $10 \div 2 = $ _____

11. $\dfrac{10}{10} = $ _____

12. $\dfrac{4}{2} = $ _____

Multiply.

13.   5
     × 3

14.   3
     × 3

15. 6 · 3 = _____

16. 3 × 4 = _____

17. Ethan bought 10 quarts of oil for his car. How many pints of oil did he buy?

18. Bailey measured the path. It is eight yards long. How many feet long is the path?

19. Caleb has $40 to spend on gifts. If he spends $10 for each gift, how many gifts can he buy?

20. Meredith drew a rectangle that measured seven inches by nine inches. What was the area of the rectangle?

Divide.

1. $5 \overline{)15}$

2. $3 \overline{)18}$

3. $3 \overline{)12}$

4. $5 \overline{)45}$

5. $5 \overline{)30}$

6. $5 \overline{)25}$

7. $3 \overline{)27}$

8. $3 \overline{)6}$

9. $14 \div 2 =$ _____

10. $50 \div 10 =$ _____

11. $\dfrac{24}{3} =$ _____

12. $\dfrac{9}{3} =$ _____

Solve for the unknown.

13. $3X = 18$

14. $7R = 21$

15. $5Y = 20$

Add. Regroup when necessary.

16.
$$\begin{array}{r} 2\,5 \\ +\,3\,8 \\ \hline \end{array}$$

17.
$$\begin{array}{r} 4\,7 \\ +\,7\,3 \\ \hline \end{array}$$

18.
$$\begin{array}{r} 6\,4 \\ +\,1\,9 \\ \hline \end{array}$$

19. What is the area of a rectangle that measures five miles by seven miles?

20. Janna bought 75 red beads and 69 blue beads for a craft project. How many beads did she buy?

Solve for the unknown.

1. $9A = 81$

2. $9X = 18$

3. $9Q = 45$

4. $9T = 72$

5. $9X = 54$

6. $9R = 27$

7. $9B = 63$

8. $9Y = 36$

Divide.

9. $35 \div 5 =$ _____

10. $24 \div 3 =$ _____

11. $\dfrac{5}{5} =$ _____

12. $\dfrac{18}{3} =$ _____

13. $3\overline{)21}$ 14. $5\overline{)25}$

15. $3\overline{)15}$ 16. $2\overline{)16}$

17. Write the symbol for parallel.

Draw a pair of lines that look parallel.

18. Write the symbol for perpendicular.

Draw a pair of lines that look perpendicular.

19. A tree is 30 feet tall. How many yards tall is the tree?

20. Cassie spent $15 for a book and $26 for a gift. How much money did she spend in all?

6

Divide.

1. $9\overline{)18}$

2. $9\overline{)36}$

3. $9\overline{)45}$

4. $9\overline{)81}$

5. $9\overline{)27}$

6. $9\overline{)54}$

7. $9\overline{)72}$

8. $9\overline{)63}$

9. $90 \div 10 =$ _____

10. $10 \div 2 =$ _____

11. $\dfrac{15}{3} =$ _____

12. $\dfrac{12}{3} =$ _____

13. $24 \div 3 =$ _____

14. $20 \div 2 =$ _____

15. $\dfrac{9}{9} =$ _____

16. $\dfrac{21}{3} =$ _____

Subtract.

17.
$$91 \\ -76$$

18.
$$42 \\ -13$$

19.
$$80 \\ -35$$

20. Hannah made 18 treats yesterday and 17 more today. She plans to divide the treats evenly among her five friends. How many treats will each friend receive?

Divide.

1. $2\overline{)4}$

2. $3\overline{)24}$

3. $9\overline{)45}$

4. $1\overline{)2}$

5. $3\overline{)18}$

6. $5\overline{)30}$

7. $9\overline{)81}$

8. $10\overline{)60}$

9. $16 \div 2 = $ _____

10. $21 \div 3 = $ _____

11. $\dfrac{63}{9} = $ _____

12. $\dfrac{5}{5} = $ _____

13. $27 \div 9 = $ _____

14. $10 \div 2 = $ _____

15. $\dfrac{40}{10} = $ _____

16. $\dfrac{12}{3} = $ _____

Divide.

17. $9\overline{)45}$

18. $2\overline{)14}$

19. $5\overline{)25}$

20. $2\overline{)18}$

21. $5\overline{)50}$

22. $3\overline{)9}$

23. $9\overline{)9}$

24. $3\overline{)27}$

25. $90 \div 10 =$ _____

26. $15 \div 3 =$ _____

27. $\dfrac{12}{2} =$ _____

28. $\dfrac{4}{1} =$ _____

29. $40 \div 5 =$ _____

30. $54 \div 9 =$ _____

31. $\dfrac{70}{10} =$ _____

32. $\dfrac{15}{5} =$ _____

Divide.

33. $5 \overline{\smash{\big)}\ 45}$

34. $3 \overline{\smash{\big)}\ 30}$

35. $1 \overline{\smash{\big)}\ 5}$

36. $10 \overline{\smash{\big)}\ 100}$

37. $1 \overline{\smash{\big)}\ 9}$

38. $5 \overline{\smash{\big)}\ 20}$

39. $9 \overline{\smash{\big)}\ 72}$

40. $2 \overline{\smash{\big)}\ 20}$

41. $6 \div 3 = \underline{\hphantom{000}}$

42. $10 \div 5 = \underline{\hphantom{000}}$

43. $\dfrac{18}{9} = \underline{\hphantom{000}}$

44. $\dfrac{35}{5} = \underline{\hphantom{000}}$

45. $8 \div 2 = \underline{\hphantom{000}}$

46. $80 \div 10 = \underline{\hphantom{000}}$

47. $\dfrac{8}{1} = \underline{\hphantom{000}}$

48. $\dfrac{3}{3} = \underline{\hphantom{000}}$

Add or subtract.

49. $\begin{array}{r} 5\ 6 \\ +\ 3\ 9 \\ \hline \end{array}$ 50. $\begin{array}{r} 6\ 2 \\ -\ 2\ 5 \\ \hline \end{array}$

51. $\begin{array}{r} 8\ 1 \\ -\ 4\ 6 \\ \hline \end{array}$

52. Draw a pair of lines that look parallel.

53. Draw a pair of lines that look perpendicular.

54. What is the area of a rectangle that measures eight feet by nine feet?

55. How many feet are in five yards?

56. How many yards are in 27 feet?

57. How many pints are in six quarts?

58. How many quarts are in 24 pints?

7

Find the area.

1.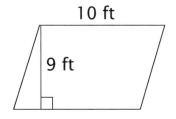

A = _____

2.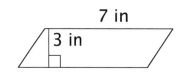

A = _____

Divide.

3. $9\overline{)36}$

4. $3\overline{)18}$

5. $5\overline{)20}$

6. $2\overline{)12}$

7. $81 \div 9 =$ _____

8. $40 \div 10 =$ _____

9. $\dfrac{4}{2} =$ _____

10. $\dfrac{63}{9} =$ _____

Solve for the unknown.

11. $6T = 12$

12. $6Y = 30$

13. $6A = 54$

14. $6F = 18$

15. $6X = 42$

16. $6R = 48$

17. $6B = 24$

18. $6Y = 36$

19. Abby raises white rabbits. She had 25 rabbits and sold 16 of them. She wants to put her remaining rabbits in cages with three rabbits in each cage. How many cages does she need?

20. Rachel bought 18 pint jars. How many quarts of her special homemade jam will fill the jars?

Divide.

1. $6\overline{)12}$

2. $6\overline{)24}$

3. $6\overline{)54}$

4. $6\overline{)30}$

5. $6\overline{)42}$

6. $6\overline{)48}$

7. $6\overline{)18}$

8. $6\overline{)36}$

9. $72 \div 9 =$ _____

10. $20 \div 5 =$ _____

11. $\dfrac{8}{2} =$ _____

12. $\dfrac{27}{3} =$ _____

Add or subtract.

13.
$$\begin{array}{r} 2\,3 \\ -\ \ 5 \\ \hline \end{array}$$

14.
$$\begin{array}{r} 7\,2 \\ +\,1\,9 \\ \hline \end{array}$$

15.
$$\begin{array}{r} 5\,3 \\ -\,4\,5 \\ \hline \end{array}$$

Multiply.

16.
$$\begin{array}{r} 2\,2 \\ \times\,1\,3 \\ \hline \end{array}$$

17.
$$\begin{array}{r} 4\,5 \\ \times\,2\,4 \\ \hline \end{array}$$

18.
$$\begin{array}{r} 1\,6 \\ \times\,3\,7 \\ \hline \end{array}$$

19. Jeremy was bored. He counted people's feet as they walked by. If he counted 20 feet, how many people had walked by?

20. A parallelogram has an area of 36 square feet. The height is six feet. What is the length of the base?

Find the area.

1. 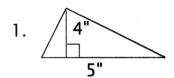   A = _____

2. 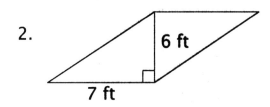   A = _____

3. A = _____

Divide.

4. 10 ⟌ 40   5. 3 ⟌ 12

6. 8 ÷ 2 = _____   7. $\frac{45}{5}$ = _____

Solve for the **unknown**.

8. 4T = 24   9. 4Y = 32

10. 4A = 16   11. 4F = 28

Multiply.

12.
```
  8 4
× 2 2
```

13.
```
  4 3
× 3 5
```

14.
```
  6 7
× 5 4
```

Add. Make 10 when possible.

15.
```
  2 5
  1 5
  2 4
+ 6 1
```

16.
```
  4 4
  3 8
  6 2
  5 6
+ 1 1
```

17.
```
  9 0
  2 3
  5 7
  1 8
+ 8 2
```

18. Can a triangle have two parallel sides?

19. Timothy counted animals at the fair. He saw 25 cows, 16 horses, 18 pigs, and 32 sheep. How many animals did he see in all?

20. Each of the animals Timothy saw (#19) had four hooves. How many hooves did he see altogether?

10

Divide.

1. $4\overline{)8}$

2. $4\overline{)32}$

3. $4\overline{)16}$

4. $4\overline{)12}$

5. $28 \div 4 =$ _____

6. $20 \div 4 =$ _____

7. $\dfrac{36}{4} =$ _____

8. $\dfrac{24}{4} =$ _____

9. $18 \div 6 =$ _____

10. $42 \div 6 =$ _____

11. $\dfrac{30}{6} =$ _____

12. $\dfrac{48}{6} =$ _____

Follow the signs.

13.　　7 1
　　　　3 4
　　　　5 9
　　　+ 2 6
　　　―――――

14.　　　6 5
　　　− 3 9
　　　―――――

15.　　8 4
　　× 6 2
　　―――――

Find the area.

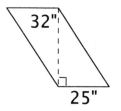

height = 5 ft

4 ft

16.　A = _____

32"

25"

17.　A = _____

2 yd

18.　　5 yd

A = _____

19.　Bria brought 24 quarts of lemonade to the picnic. How many gallons did she bring?

20.　Alaina has 40 quarters in her piggy bank. How many dollars does she have?

Find the average of the given numbers.

    1. 2, 7, 8, 3   average = _____

    2. 4, 3, 6, 7, 10   average = _____

    3. 5, 11, 14   average = _____

Solve for the unknown.

    4. $7A = 49$ 5. $7X = 42$

    6. $7B = 56$ 7. $8D = 64$

    8. $8Q = 56$ 9. $8G = 72$

Divide.

    10. $9\overline{)63}$ 11. $6\overline{)48}$

    12. $5\overline{)40}$ 13. $21 \div 3 =$ _____

    14. $32 \div 4 =$ _____   15. $\dfrac{80}{10} =$ _____

Find the area.

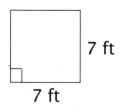

7 ft

7 ft

1 mi

8 mi

16. A = _____

17. A = _____

18. Can a triangle have a pair of sides that are perpendicular to each other?

19. Sarah bought 11 feet of red ribbon and 16 feet of yellow ribbon. How many *yards* of ribbon did she buy altogether?

20. Alexis is trying to drink more water. She drank three pints on Monday, two pints on Tuesday, four pints on Wednesday, and three pints on Thursday. What is the average daily number of pints she drank?

   How many quarts of water did she drink in all?

Divide.

1. $7\overline{)28}$

2. $7\overline{)63}$

3. $8\overline{)56}$

4. $8\overline{)16}$

5. $7\overline{)14}$

6. $7\overline{)35}$

7. $8\overline{)24}$

8. $8\overline{)72}$

9. $48 \div 8 =$ _____

10. $42 \div 7 =$ _____

11. $\dfrac{21}{7} =$ _____

12. $\dfrac{40}{8} =$ _____

13. $49 \div 7 =$ _____

14. $32 \div 8 =$ _____

15. $\dfrac{64}{8} =$ _____

16. $\dfrac{56}{7} =$ _____

Find the area.

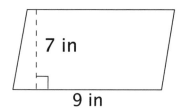

7 in

9 in

17. A = _____

18. Which bag of apples weighs more: the three-pound bag or the 50-ounce bag?

19. Aiden gave me 28 quarters. How many dollars did I receive?

20. Find the average of the numbers: 1, 2, 7, 10, 12, 16

Divide.

1. $4\overline{)36}$

2. $6\overline{)24}$

3. $8\overline{)32}$

4. $7\overline{)49}$

5. $8\overline{)56}$

6. $4\overline{)20}$

7. $6\overline{)42}$

8. $6\overline{)60}$

9. $16 \div 8 = \underline{\hspace{1cm}}$

10. $35 \div 7 = \underline{\hspace{1cm}}$

11. $\dfrac{48}{8} = \underline{\hspace{1cm}}$

12. $\dfrac{7}{7} = \underline{\hspace{1cm}}$

13. $12 \div 4 = \underline{\hspace{1cm}}$

14. $36 \div 6 = \underline{\hspace{1cm}}$

15. $\dfrac{70}{7} = \underline{\hspace{1cm}}$

16. $\dfrac{12}{6} = \underline{\hspace{1cm}}$

Divide.

17. $8\overline{)8}$

18. $7\overline{)56}$

19. $4\overline{)28}$

20. $7\overline{)42}$

21. $8\overline{)24}$

22. $4\overline{)32}$

23. $8\overline{)80}$

24. $7\overline{)21}$

25. $18 \div 6 = $ _____

26. $4 \div 4 = $ _____

27. $\dfrac{14}{7} = $ _____

28. $\dfrac{72}{8} = $ _____

29. $40 \div 4 = $ _____

30. $64 \div 8 = $ _____

31. $\dfrac{6}{6} = $ _____

32. $\dfrac{48}{6} = $ _____

33. 8⟌40

34. 4⟌24

35. 7⟌63

36. 6⟌30

37. 7⟌28

38. 4⟌8

39. 6⟌54

40. 4⟌16

Follow the signs.

41.　　3 5
　　　　7 2
　　　　1 5
　　　+4 8

42.　　9 1
　　　−3 6

43.　　7 5
　　　×5 8

44. What is the area of a triangle with a base of seven yards and a height of two yards?

45. Find the average of the numbers: 5, 9, 13

46. How many quarts are in six gallons?

47. How many gallons are in 32 quarts?

48. How many quarters are in $9?

49. How many dollars are equal to 20 quarters?

50. How many ounces are in two pounds?

Find the area of the figures.

1.

8 in

6 in

12 in

A = _____

2.

2 ft

7 ft

A = _____

3.

3 ft

8 ft

9 ft

A = _____

4.

52 in

36 in

A = _____

Multiply using the shortcut method.

5. 14 × 20 = _____

6. 22 × 30 = _____

7. 43 × 30 = _____

8. 51 × 20 = _____

Divide.

9. $42 \div 7 =$ _____

10. $100 \div 10 =$ _____

11. $64 \div 8 =$ _____

12. $27 \div 9 =$ _____

Add or subtract.

13.
$$\begin{array}{r} 352 \\ + 126 \\ \hline \end{array}$$

14.
$$\begin{array}{r} 811 \\ - 349 \\ \hline \end{array}$$

15.
$$\begin{array}{r} 607 \\ + 785 \\ \hline \end{array}$$

Fill in the blanks.

16. 8 qt = _____ gal

17. 28 quarters = $_____

18. 27 ft = _____ yd

19. A 50-foot row of corn needs to be weeded. Five people each weeded the same amount. How many feet of corn did each person weed?

20. Judah needs $315 to buy the bicycle he wants. He has saved $227. How much more money does he need to save?

Follow the directions.

1. Write using numbers: 200,000 + 20,000 + 1,000 + 300 + 40 + 6

_____

2. Write using numbers: 3,000,000 + 400,000 + 60,000 + 7,000

_____

3. Write using place-value notation: 6,123,500

_____

4. Write using place-value notation: 4,500,000

_____

Find the area of the figures.

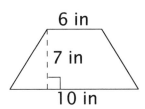

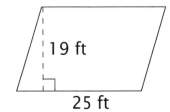

5. A = _____

6. A = _____

Multiply.

7. $6 \times 200 =$ _____

8. $24 \times 200 =$ _____

9. $17 \times 100 =$ _____   10. $32 \times 200 =$ _____

Divide.

11. $56 \div 8 =$ ____   12. $49 \div 7 =$ ____

13. $9 \div 9 =$ ____   14. $24 \div 6 =$ ____

15. How many parallel lines does a trapezoid have?

16. How many ounces are in a five-pound bag of flour?

17. How many square inches are in a triangle with a base of eight inches and a height of two inches?

18. Willow bought 16 pints of ice cream. How many quarts of ice cream does she have?

     How many gallons of ice cream does she have?

Follow the directions.

1. Write in standard notation and read the number:

   $7 \times 1{,}000{,}000{,}000 + 6 \times 100{,}000{,}000 + 3 \times 10{,}000{,}000 + 2 \times 1{,}000{,}000 + 4 \times 100{,}000$ _____

2. Write in standard notation and read the number:

   $5 \times 100{,}000{,}000 + 5 \times 10{,}000{,}000 + 5 \times 1{,}000{,}000 + 4 \times 100{,}000 + 3 \times 10{,}000 + 1 \times 1{,}000$ _____

3. Write in standard notation:

   one trillion, six hundred thirty-five billion, seven hundred twenty-one million _____

4. Write in standard notation:

   four million, three hundred fifteen thousand, twenty-one

   _____

5. Write using expanded notation: 8,250,000,000

   _____

6. Write using expanded notation: 3,400,000,000,274

   _____

Multiply.

7. $10 \times 60 =$ _____

8. $12 \times 40 =$ _____

9. $20 \times 200 =$ _____

Add or subtract.

10. $\begin{array}{r} 2{,}543 \\ +\ 8{,}067 \\ \hline \end{array}$

11. $\begin{array}{r} 6{,}460 \\ -\ 192 \\ \hline \end{array}$

12. $\begin{array}{r} 1{,}247 \\ 3{,}598 \\ +\ 6{,}013 \\ \hline \end{array}$

Divide.

13. $63 \div 9 =$ _____

14. $36 \div 6 =$ _____

15. $21 \div 7 =$ _____

16. $8 \div 8 =$ _____

17. Find the average of the numbers: 3, 5, 11, 13

18. Mackenzie spent 16 hours a month babysitting. How many hours did she babysit in 12 months?

Divide.

1. 3 | 10

2. 6 | 25

3. 9 | 30

4. 4 | 31

5. 2 | 15

6. 5 | 47

7. 8 | 20

8. 7 | 41

9. 8 | 37

Add or subtract.

10.   3,076
      − 1,467

11.   4,654
      − 3,298

12.   6,512
      + 7,285

Multiply.

13. $11 \times 700 = \underline{\hspace{1cm}}$

14. $12 \times 300 = \underline{\hspace{1cm}}$

15. $21 \times 400 = \underline{\hspace{1cm}}$

16. Write using standard notation:

    eight billion, three hundred ten million, six hundred seventy-five thousand, four hundred twenty _____

17. A pet shop owner has 18 canaries. Each each of his bird cages can hold four birds. How many full cages will he have?

    How many birds will be left over for another cage?

18. What is the area of a rectangle with a base of 45 feet and a height of 38 feet?

Rewrite each problem using place-value notation. Then multiply and compare your answers.

1. 252
    × 4   × _____

2.   4
    × 252   × _____

Divide.

3. 3 | 90

4. 6 | 240

5. 2 | 120

6. 5 | 200

7. 6 | 55

8. 8 | 26

9. 9 | 64

10. 7 | 30

Fill in the blanks.

11. 3 lb = _____ oz

12. 4 gal = _____ qt

13. 33 ft = _____ yd

14. 1 ton = _____ lb

15. 6 tons = _____ lb

16. 3 tons = _____ lb

17. Jacob wants to divide 31 dollar bills among his four children.
     How many will each child receive?

     How many dollar bills will Jacob have left over?

18. Write in standard notation:

     five trillion, four hundred billion, six hundred million, five

     _____

Divide. Check your work by multiplying upside down.

1. $2\overline{)42}$

2. $9\overline{)92}$

3. $3\overline{)67}$

4. $4\overline{)19}$

5. $5\overline{)21}$

6. $8\overline{)480}$

Write the numbers in columns and add.

7. $25 + 75 + 45 =$ _____

8. $8 + 7 + 4 + 3 + 2 =$ _____

9. $95 + 345 =$ _____

Fill in the blanks.

10. 4 tons = _____ lb

11. $31 = _____ quarters

12. 10 lb = _____ oz

13. 1 mile = _____ feet

14. 3 mi = _____ ft

15. 5 mi = _____ ft

16. Find the average of the numbers: 4, 8, 15, 25

17. Taylor has 25 pictures to put in his photo album. He can fit four pictures on a page. How many pages will he fill?

How many pictures will he have left over?

18. How many feet are in five yards?

LESSON TEST

Divide. Check your work by multiplying.

1. $3\overline{)714}$

2. check for #1

3. $5\overline{)628}$

4. check for #3

5. $4\overline{)368}$

6. check for #5

7. $7\overline{)113}$

8. check for #7

Multiply.

9. 453
    × 46

10. 839
    × 25

11. 851
    × 69

Fill in the blanks.

12. 40 yd = _____ ft

13. 2 mi = _____ ft

14. 5 lb = _____ oz

15. Write in standard notation:

    2 × 1,000,000,000 + 4 × 100,000,000 + 9 × 10,000,000 +
    5 × 1,000,000

    _____

16. Do the lines appear to be parallel or perpendicular?

Divide. Include a fraction in your answer if the problem does not divide evenly. Multiply to check your work.

1. $5\overline{\smash{)}895}$

2. check for #1

3. $8\overline{\smash{)}356}$

4. check for #3

5. $3\overline{\smash{)}614}$

6. check for #5

7. $6\overline{\smash{)}578}$

8. check for #7

9. Driving at a steady speed, Emma covered 300 miles in 5 hours. How many miles did Emma drive each hour?

10. How many feet are in three miles?

11. The area of a parallelogram is 35 square yards. The base is seven yards. What is the height?

12. Change the base and height of the parallelogram in #11 to feet and multiply to find the area in square feet.

13. Which weighs more, eight tons or 1,600 pounds?

14. How many perpendicular corners does a rectangle have?

15. Richard was born in May of 1974. How old was he on his birthday in 2003?

Fill in the blanks.

1. 29 to the nearest ten is _____.

2. 109 to the nearest hundred is _____.

3. 1,168 to the nearest thousand is _____.

4. 14 to the nearest ten is _____.

5. 355 to the nearest hundred is _____.

6. 4,500 to the nearest thousand is _____.

Estimate the answer. Then divide and compare your answers.

7. $6\overline{)869}$ → $6\overline{)(\qquad)}$ 8. $6\overline{)869}$

Divide and check by multiplying.

9. $2\overline{)652}$

10. check for #9

11. $7\overline{)239}$

12. check for #11

13. Abigail walked two miles. How many feet did she walk?

14. How many pounds are in five tons?

15. Thomas drove 500 miles every day for five days. How many miles did he drive in all?

16. What is the area of a trapezoid with bases of 13 and 17 feet and a height of 11 feet?

Find the area of the trapezoids.

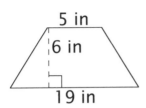

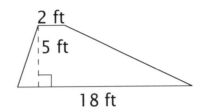

1. A = _____

2. A = _____

Follow the directions.

3. Write in standard notation:

    $3 \times 1{,}000{,}000{,}000 + 7 \times 100{,}000{,}000 + 6 \times 10{,}000{,}000 + 1 \times 1{,}000{,}000 + 8 \times 100{,}000$

    _____

4. Write in standard notation:

    two trillion, four hundred thirteen billion, two hundred eighty-three million, three hundred fifty thousand

    _____

5. Write using expanded notation: 75,123,000

    _____

Multiply.

6. $28 \times 60 =$ _____

7. $13 \times 400 =$ _____

8. $56 \times 700 =$ _____

Divide. Write your remainders without using fractions.

9. $2 \overline{)60}$

10. $8 \overline{)65}$

11. $4 \overline{)39}$

12. $3 \overline{)669}$

13. $5 \overline{)257}$

14. $8 \overline{)829}$

Fill in the blanks.

15. 31 to the nearest ten is _____ .

16. 249 to the nearest hundred is _____ .

17. 2,503 to the nearest thousand is _____ .

Estimate the answer. Divide and compare your answers. Include a fraction in your answer if the problem does not divide evenly.

18. $5 \overline{)908}$ → $5 \overline{)(\quad\quad)}$

19. $5 \overline{)908}$

Divide. Include a fraction in your answer if the problem does not divide evenly. Multiply to check your work.

20. $4 \overline{)865}$

21. check for #20

22. $2 \overline{)157}$

23. check for #22

24. How many feet are in five miles?

25. How many pounds are in nine tons?

26. Five hundred and sixty-eight ears listened to Mr. Demme's speech. How many people were listening?

Divide. Include a fraction in your answer if the problem does not divide evenly. Multiply to check your work.

1. 34⟌517

2. check for #1

3. 18⟌367

4. check for #3

5. 5⟌78

6. check for #5

7. 9⟌934

8. check for #7

Find the area of the figures.

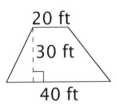

20 ft
30 ft
40 ft

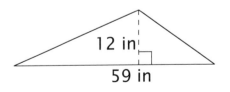

12 in
59 in

9. _____

10. _____

Find the area of the figure.

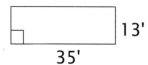

13'

35'

11. _____

Fill in the blanks.

12. 144 in = _____ ft

13. 20 feet = _____ inches

14. 72 in = _____ ft

15. Jaden has 48 quarters. How many dollars is that?

16. Logan drove at 55 miles an hour for 330 miles. Divide to find the number of hours he drove.

17. On Logan's 330-mile trip, he used 11 gallons of gasoline. How many miles did he travel for each gallon he used?

18. Which is greater: 7 tons or 140,000 pounds?

Divide. Include a fraction in your answer if the problem does not divide evenly. Multiply to check your work.

1. 5 | 4 5 6 2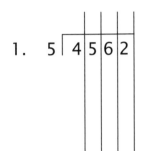

2. check for #1

3. 7 | 1 4 9 3

4. check for #3

5. 82 | 3 5 8

6. check for #5

7. 21 | 8 4 0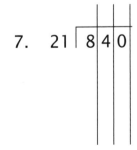

8. check for #7

Multiply.

9.

4	1	3	7
×		1	3

10.

2	4	2	8
×		7	5

11.

7	8	0	1
×		3	6

12. How many feet are in 396 inches?

13. How many parallel lines are in an uppercase "E"?

14. Cameron wants to hang a picture that weighs 80 ounces. The label on the package says the picture hanger will hold 10 pounds. Is the hanger strong enough to hold his picture?

15. A pool of water held 2,272 gallons. How many quarts of water did the pool hold?

Divide. Include a fraction in your answer if the problem does not divide evenly. Multiply to check your work.

1. 25 ⟌ 2 7 2 4

2. check for #1

3. 81 ⟌ 2 3 3 4

4. check for #3

5. 6 ⟌ 8 6 7 3

6. check for #5

7. 12 ⟌ 3 7 2

8. check for #7

Multiply upside down.

9.

		3	4	5
×	1	4	7	2

10.

		8	3	7
×	4	7	5	4

11.

		6	3	2
×	3	5	8	1

12. What is the area of a trapezoid with bases of 13 feet and 19 feet and a height of 41 feet?

13. Thirty-four people want to go to the picnic. All the available cars hold five people each. How many cars are needed to drive everyone who wants to go to the picnic?

14. Draw two lines that look parallel.

Divide. Include a fraction in your answer if the problem does not divide evenly. Multiply to check your work.

1. 29 ⟌ 3 3 4 5 6
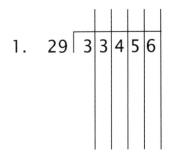

2. check for #1

3. 180 ⟌ 3 7 4 6 7 5

4. check for #3

5. 7 ⟌ 4 5 9 2

6. check for #5

7. 1 5 ⟌ 1 6 3 1
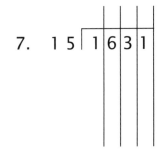

8. check for #7

Find the missing part of each parallelogram or rectangle.

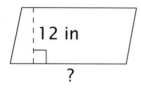

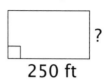

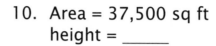

9. Area = 336 sq in
    base = _____

10. Area = 37,500 sq ft
    height = _____

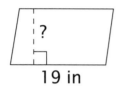

11. Area = 323 sq in
    height = _____

12. Riley weighs 56 pounds. How many ounces does she weigh?

13. Tyler drove 270 miles in 6 hours. How many miles did he drive in one hour?

14. Find the average weight: 780 pounds; 4,670 pounds; 550 pounds.

    What is the average weight in tons?

Find the volume.

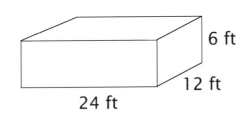

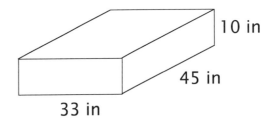

1. V = _____ cubic feet

2. V = _____ cubic inches

Divide and check by multiplying. Use estimation to help you if needed.

3. 5 | 2 7 3 7 0

4. check for #3

5. 32 | 8 2 2 6 1

6. check for #5

Fill in the blanks.

7. 28 pt = _____ qt   8. $44 = _____ quarters

9. 48 oz = _____ lb   10. 6 tons = _____ lb

11. 100 qt = _____ gal   12. 60 in = _____ ft

13. A rectangular pool measures 20 feet by 15 feet by 6 feet. How many cubic feet of water would fill the pool to the brim?

14. How many gallons of water are needed to fill the pool in #13?

15. A gallon of water weighs eight pounds. What will the water in #13 weigh when the pool is filled to the brim?

Solve.

1. $\dfrac{4}{7}$ of $21 =$ _____

2. $\dfrac{2}{3}$ of $9 =$ _____

3. $\dfrac{1}{9}$ of $36 =$ _____

4. $\dfrac{3}{5}$ of $75 =$ _____

5. $\dfrac{5}{6}$ of $18 =$ _____

6. $\dfrac{1}{2}$ of $28 =$ _____

Divide and check by multiplying. Use estimation to help you if needed.

7. $3 \overline{\smash{)}2\,9\,1\,4\,7}$

8. check for #7

9. $46 \overline{\smash{)}4\,6\,5\,3\,6}$

10. check for #9

Fill in the blanks.

11. 36 ft = _____ yd

12. 24 quarters = $_____

13. 8 mi = _____ ft

14. Isabella collected 32 flowers for a nature project. She decided to press and keep 1/4 of them. How many flowers did she keep?

_____

15. What is the volume of a rectangular shape that measures 100 inches by 100 inches by 90 inches?

_____

Write the Arabic numeral represented by each Roman numeral.

1. XIV = _____

2. LXXII = _____

3. CCXXX = _____

4. XCIX = _____

Write the Roman numeral represented by each Arabic numeral.

5. 41 = _____

6. 85 = _____

7. 333 = _____

8. 29 = _____

Solve.

9. $\frac{1}{2}$ of 12 = _____

10. $\frac{3}{8}$ of 64 = _____

11. $\frac{2}{5}$ of 15 = _____

Find the area.

11 ft

13 ft

25 ft

12. A = _____

Divide and check by multiplying. Use estimation to help you if needed.

13. 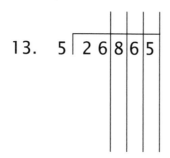 5 ⟌ 2 6 8 6 5

14. check for #13

15. 216 ⟌ 4 9 3 4 8 4

16. check for #15

17. Find the average of the numbers: 16, 43, 58, 91

18. Morgan spotted 24 horses grazing in a field. One sixth of the horses were black. How many horses were black?

Find the denominator and numerator of each fraction.

1. $\dfrac{\text{numerator}}{\text{denominator}} = \underline{\hspace{1cm}}$

2. $\dfrac{\text{numerator}}{\text{denominator}} = \underline{\hspace{1cm}}$

Shade the rectangles to show the given fractions.

3. $\dfrac{\text{numerator}}{\text{denominator}} = \dfrac{3}{4}$

4. $\dfrac{\text{numerator}}{\text{denominator}} = \dfrac{2}{5}$

Write the Arabic numeral represented by each Roman numeral.

5. XXVI = _____

6. XLIII = _____

7. CLXV = _____

8. CXCII = _____

Write the Roman numeral represented by each Arabic numeral.

9. 47 = _____

10. 18 = _____

11. 219 = _____

12. 154 = _____

Divide and check by multiplying. Use estimation to help you if needed.

13. 31 | 8 7 9 3

14. check for #13

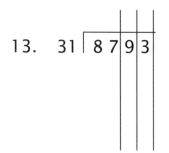

15. 14 | 1 9 8 4 5

16. check for #15

17. What is the area of a triangle with a base of 20 inches and a height of 9 inches?

18. Paige drove 165 miles at 55 miles an hour. How many hours did she spend driving?

Write the Arabic numeral represented by each Roman numeral.

1. MMCC = _____

2. DXXV = _____

3. DCCL = _____

4. CMXXIX = _____

Write the Roman numeral represented by each Arabic numeral.

5. 58 = _____

6. 520 = _____

7. 3,700 = _____

8. 1965 = _____

Find the denominators and numerators of the fractions.

9. $\dfrac{\text{numerator}}{\text{denominator}}$ = ___

10. $\dfrac{\text{numerator}}{\text{denominator}}$ = ___

Divide and check by multiplying. Use estimation to help you if needed.

11. 64 | 8 1 | 3 | 3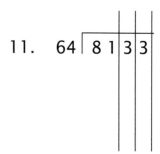

12. check for #11

13. 500 | 6 4 | 3 | 0 | 0 | 0

14. check for #13

15. What is the area of a trapezoid with bases of seven feet and nine feet and a height of eight feet?

16. Round 561 to the nearest hundred.

17. At an average speed of 15 miles an hour, how long would it take to travel 135 miles?

18. Mom bought 25 pounds of apples. One fifth of the apples spoiled before she could use them. How many pounds of unspoiled apples does she have left?

Divide. Include a fraction in your answer if the problem does not divide evenly. Multiply to check your work.

1. 13 ⟌ 568

2. check for #1

3. 30 ⟌ 971

4. check for #3

5. 5 ⟌ 4 5 1 5 4

6. check for #5

7. 24 ⟌ 6 8 9 3 2

8. check for #7

Find the volume.

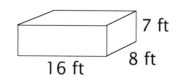

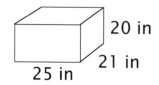

9. V = _____

10. V = _____

Solve.

11. $\dfrac{1}{2}$ of 32 = _____

12. $\dfrac{5}{8}$ of 64 = _____

13. $\dfrac{5}{6}$ of 42 = _____

Find the denominators and numerators of the fractions.

14. $\dfrac{\text{numerator}}{\text{denominator}}$ = ____

15. $\dfrac{\text{numerator}}{\text{denominator}}$ = ____

16. How many inches long is an eight-foot board?

17. What Arabic numeral is represented by the Roman numeral XCIX?

18. Write 2,453 with Roman numerals.

Divide. Write your remainders without using fractions.

1. 4 ⟌ 80

2. 7 ⟌ 53

3. 8 ⟌ 648

4. 5 ⟌ 396

Divide. Include a fraction in your answer if the problem does not divide evenly. Check your answers.

5. 25 ⟌ 631

6. check for #5

7. 16 ⟌ 349

8. check for #7

9. 6 ⟌ 3 0 | 4 | 5 | 8

10. check for #9

11. 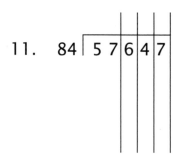 84⟌5 7 6 4 7

12. check for #11

Find the area of each figure.

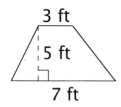

3 ft
5 ft
7 ft

13. A = _____

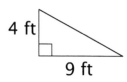

4 ft
9 ft

14. A = _____

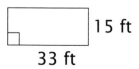

10 in
25 in

15. A = _____

15 ft
33 ft

16. A = _____

Find the volume.

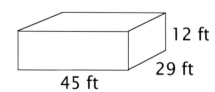

12 ft
29 ft
45 ft

17. V = _____

Fill in the blanks.

18. 27 ft = _____ yd

19. 40 pt = _____ qt

20. 20 qt = _____ gal

21. 5 dollars = _____ quarters

22. 4 lb = _____ oz

23. 1 mi = _____ ft

24. 5 tons = _____ lb

25. 36 in = _____ ft

26. 40 ft = _____ in

27. 49 to the nearest ten is _____ .

28. 4,009 to the nearest thousand is _____ .

29. 459 to the nearest hundred is _____ .

Solve.

30. $\frac{1}{3}$ of 12 = _____

31. $\frac{3}{7}$ of 21 = _____

32. $\frac{5}{8}$ of 32 = _____

Find the denominators and numerators of the fractions.

33. $\frac{\text{numerator}}{\text{denominator}}$ = _____

34. $\frac{\text{numerator}}{\text{denominator}}$ = _____

35. Write in standard notation:

$2 \times 1{,}000{,}000{,}000 + 5 \times 100{,}000{,}000 + 4 \times 10{,}000{,}000 + 3 \times 1{,}000{,}000 + 9 \times 100{,}000$

_____

36. Find the average of the numbers: 5, 12, 13, 21, 24

37. What Arabic numeral is represented by the Roman numeral MMCLVIII?

38. Write the given year with Roman numerals: 1975